WHEN DINOSAURS RULED THE EARTH

THE STEGOSAURUS

Written by Tracy Vonder Brink   Illustrated by Riley Stark

TABLE OF CONTENTS

A Crabtree Seedlings Book

Crabtree Publishing
crabtreebooks.com

School-to-Home Support for Caregivers and Teachers

This book helps children grow by letting them practice reading. Here are a few guiding questions to help the reader with building his or her comprehension skills. Possible answers appear here in red.

Before Reading:

- What do I think this book is about?
  - *I think this book is about dinosaurs.*
  - *I think this book is about the dinosaur called* Stegosaurus.
- What do I want to learn about this topic?
  - *I want to learn how long* Stegosaurus *was.*
  - *I want to learn what* Stegosaurus *ate.*

During Reading:

- I wonder why...
  - *I wonder why paleontologists study fossils.*
  - *I wonder how* Stegosaurus *used its beak.*
- What have I learned so far?
  - *I have learned that dinosaurs lived before people.*
  - *I have learned that* Stegosaurus *lived around the world.*

After Reading:

- What details did I learn about this topic?
  - *I have learned that* Stegosaurus *had spikes on its tail.*
  - *I have learned that* Stegosaurus *was a herbivore.*
- Read the book again and look for the glossary words.
  - *I see the word* ***paleontologists*** *on page 4 and the word* ***herbivore*** *on page 15. The other glossary words are found on page 22.*

STEGOSAURUS

Many dinosaurs once roamed Earth.

They lived long before people.

Some dinosaurs became **fossils** after they died.

Paleontologists study fossils to learn about dinosaurs.

Stegosaurus was a dinosaur that lived about 150 million years ago.

Stegosaurus fossils have been found around the world.

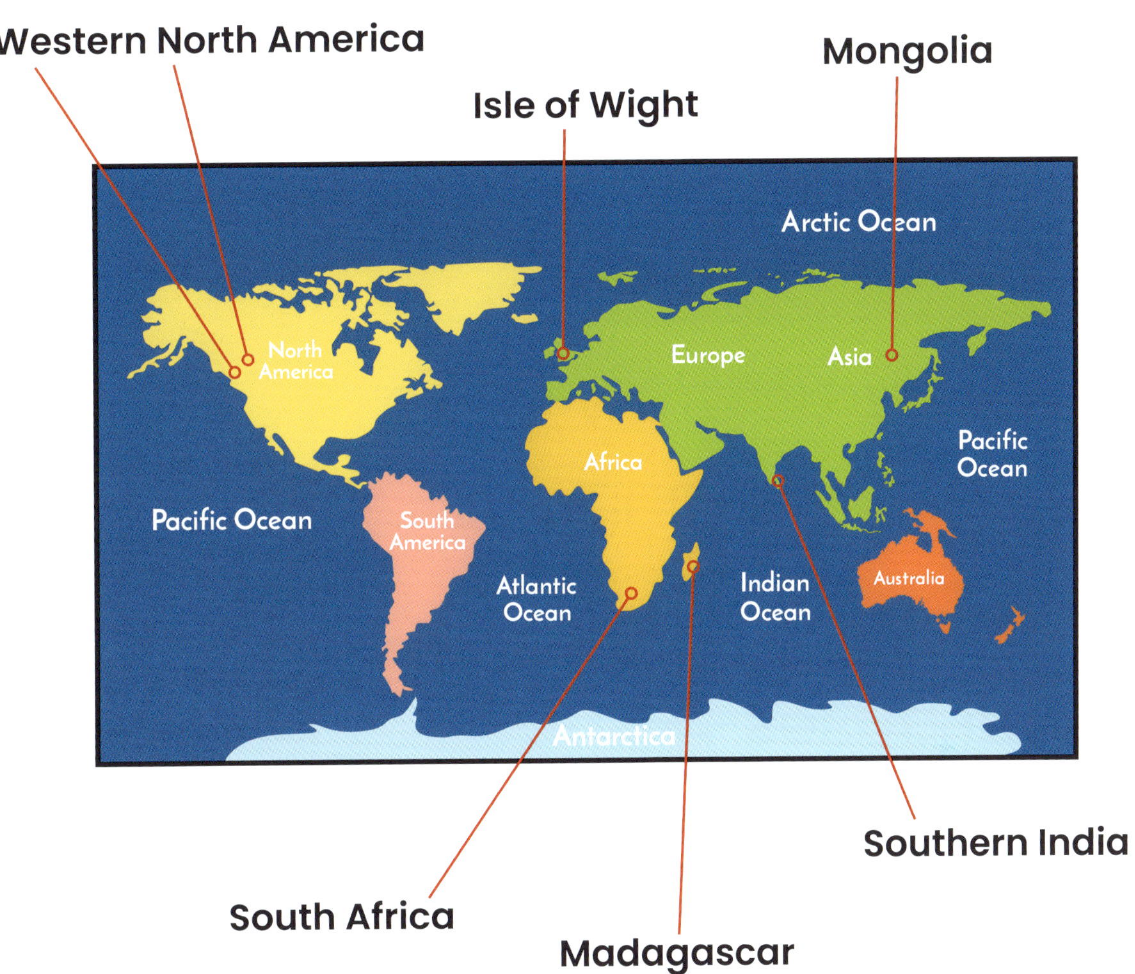

Stegosaurus *fossils have been found in North America, Europe, India, Asia and Africa.*

Stegosaurus grew up to 30 feet (9 m) long.

That's as long as some school buses.

Stegosaurus walked on four legs.

Its back legs were longer than its front legs.

Stegosaurus *probably could not move very fast because of the difference in length between its front and back legs.*

Stegosaurus had a small brain.

Its brain is believed to have been about the size and shape of a bent hot dog.

Stegosaurus *probably ate low-lying plants such as ferns and shrubs.*

Stegosaurus was a **herbivore**.

It had a beak, which it used to tear up plants to eat.

Stegosaurus had sharp spikes on its tail.

Its spiked tail may have been used for **defense**.

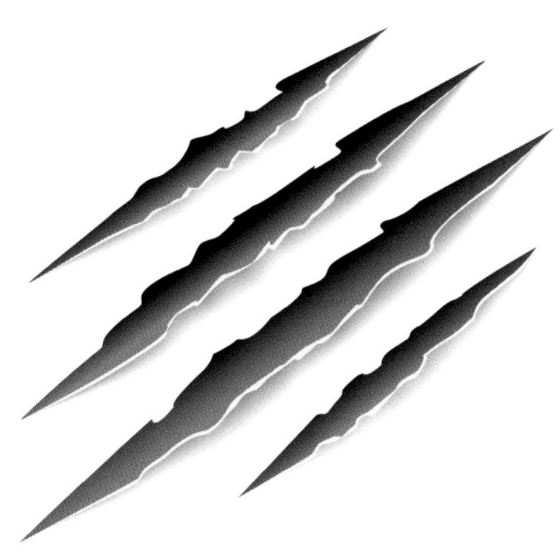

Stegosaurus had two rows of bony **plates** along its back.

The plates might also have helped Stegosaurus *control its body temperature.*

How did *Stegosaurus* use its teeth?

More fossils need to be discovered to find the answers.

Stegosaurus *had small teeth shaped like pegs.*

Glossary

attract (UH-trakt): To cause interest or attention

defense (duh-FENSE): Something used to protect from attack

fossil (FAWS-uhl): The traces, prints, or remains of plants and animals that lived long ago

herbivore (HER-bi-vor): An animal that only eats plants

paleontologist (PAY-lee-en-TAW-luh-jist): A scientist who studies fossils to learn about past life on Earth

plates (PLAYTS): Hard, flat pieces found on the bodies of certain animals

Index

About the Author

Tracy Vonder Brink loves to learn about science and nature. She has written many nonfiction books for kids and is a contributing editor for three children's science magazines. Tracy lives in Cincinnati, Ohio, with her husband, two daughters, and two rescue dogs. She would like to see *Stegosaurus* use its beak.

Written by: Tracy Vonder Brink
Illustrated by: Riley Stark
Designed by: Rhea Wallace
Series Development: James Earley
Proofreader: Melissa Boyce
Educational Consultant: Marie Lemke M.Ed.

Photographs:
Shutterstock: Engineer Studio: p. 5; Rimma R: p. 7; Danny Ye: p. 21

Crabtree Publishing

crabtreebooks.com 800-387-7650

Printed in the U.S.A./072023/CG20230214

Published in Canada
Crabtree Publishing
616 Welland Ave.
St. Catharines, Ontario
L2M 5V6

Published in the United States
Crabtree Publishing
347 Fifth Ave
Suite 1402-145
New York, NY 10016

Library and Archives Canada Cataloguing in Publication
Available at Library and Archives Canada

Library of Congress Cataloging-in-Publication Data
Available at the Library of Congress

Hardcover: 978-1-0396-9649-5
Paperback: 978-1-0396-9756-0
Ebook (pdf): 978-1-0396-9970-0
Epub: 978-1-0396-9863-5